Alachua County ARES

Operation Hurricane Test

May 6, 2017

After Action
Report/Improvement Plan

May 22, 2017

DRAFT

Mobile antenna tower getting ready to deploy at the EOC

Copyright © 2017 Gordon L. Gibby MD

ISBN: **1548062200**
ISBN-13: **978-1548062200**

CONTENTS

ACKNOWLEDGMENTS

The Alachua County ARES group would like to acknowledge all the people who helped us make this growth possible, including all our own volunteers who worked so hard for so many months, the Alachua County Emergency Manager's office, Alachua County Emergency Operations Center, the Gainesville Red Cross, the University of Florida Housing Department who allowed us to get on a dorm rooftop to repair W4DFU-7/8's antenna, the Florida Forest Service which graciously allows us to cover the western end of the county using one of their lookout towers and antennas and even provided a crank up antenna tower for this exercise.

SECTION 1: EXERCISE OVERVIEW

Exercise Details

Exercise Name
2017 ARES Hurricane Test

Type of Exercise
Full Scale Exercise

Exercise Start Date
May 6 2017

Exercise End Date
May 6 2017

Duration
4 hours

Location
4 locations: Alachua County EOC; Senior Center (Hurricane shelter); Gainesville Red Cross; Easton-Newberry Sports Complex (Hurricane shelter)

Sponsor
Alachua County ARES, a component of the American Radio Relay League (ARRL)

Program
Amateur Radio Emergency Service

Mission
Communications Support

Capabilities
VHF local communications, analog voice and digital (AX.25 packet)

HF local and national communications, analog voice and digital (PSK31, WINLINK)

Radio Email to anywhere, via WINLINK

Scenario Type

Total Communications Failures

Exercise Planning Team

Gordon L. Gibby MD KX4Z NCS521

Jeff Capehart W4UFL

Participating Organizations

Alachua County, Florida

Emergency Operations Center

Red Cross (use of their building)

Senior Center (use of their building)

Easton-Newberry Sports Complex (use of their grounds)

State

Florida Forest Service, State of Florida (gracious provision of 100 foot mobile antenna tower, as well as use of the Forest Grove Lookout Tower.

Number of Participants

- Players - 13
- Controllers - 1
- Evaluators – 3

Amateur Radio Volunteers at the Gainesville Senior Center (a hurricane shelter)

After Action Report / Improvement Plan (AAR/IP)   Alachua County ARES   2017 Hurricane Test

SECTION 2: EXERCISE DESIGN SUMMARY

Exercise Purpose and Design

The Alachua ARES Hurricane Test Full Scale Exercise was developed to give the local ARES component a very realistic simulation of the communications tasks and hurdles they might face if their skills were actually needed after a hurricane.   Previous years' testing has been somewhat minimal, so this was a tremendous step up in the "communications bar" the volunteer amateur radio operators tackled.

Training for this type of scenario – total loss or overwhelming of local conventional communications systems (telephone/Internet) began almost nine months before this exercise, as local skills, assets, and strategies began to be sharpened.

Development of this exercise began 3 months before with completion of a FEMA 120 Exercise Development course by the Exercise Developer after encouragement by Dave Welker, of the Marion County Hospital Emergency Communications group.

As the Exercise Planning was reaching completion, with a 25 page Participant Workbook and 35+ page supervisory document, local training of volunteer ARES members accelerated with group and individual training sessions, including a real-equipment, power down simulation carried out at a private residence to give participants a very similar experience to what they would face during the Full Scale Exercise.

Complicating this Full Scale Exercise was the fact that Amateur Radio emergency communications techniques and systems are not standardized nationwide.   Amateur radio is not a monolithic radio service; instead participants draw from a plethora of modalities, skills, frequencies, and procedures to communicate, from extreme low frequency ground wave, to VHF repeater, voice, digital, and even Earth-Satellite or Earth-Moon-Earth communications.   The local group leaders were both continuously training participants, while themselves learning at the same time, and developing increasingly better understanding of which techniques were optimal for the planned scenario.   A specific example of this is that a significant technique & software package (EASYTERM by UZ7HO) was discovered within the last two months of planning, and required intensive individual training of location leaders to make it possible to have as an alternate communications possibility during the Full Scale Exercise.   The net result was a much greater understanding of the best communications techniques at the end of the Full Scale Exercise, and an acute understanding of local weaknesses.

Exercise Objectives

The list below includes the objectives established for the Full Scale Exercise. As might be expected with an exercise of this complexity, being such a radical expansion of previous local amateur radio emergency communications capabilities tests, not all of the exercise objectives were even demonstrated, and there was uneven performance across the locations, modalities, and objectives.

- **Objective 1: Assess the capabilities of groups and individuals at EOC, Red Cross, up to 2 local shelter sites, and 1 local hospital, to create, manage, and position antennas in response to communications goals and weather-induced damage of existing antennas and repeaters; provided that during this exercise no person shall go onto any roof or use any slingshot or other lofting mechanism in the vicinity of any power line greater than 240VAC**

  Capability: ANTENNA PLACEMENT

- **Objective 2: Assess the capability to place an emergency simplex repeater and utilize it to provide communications between all the locations involved in the Exercise.**

  Capability: EMERGENCY SIMPLEX REPEATER

- **Objective 3: Assess the capability to utilize WINLINK text messages, ICS forms, and attachments on both VHF and HF frequencies to meet realistic emergency communications needs.**

  CAPABILITY: WINLINK COMMUNICATIONS

- **Objective 4: Assess the capability to flexibly find and employ backup power systems of any available type at all locations involved in the Exercise**

  CAPABILITY: BACKUP POWER

- **Objective 5: Assess the capability to move (when travel is "safe") to a new location and expeditiously resume communications on VHF.**

  CAPABILITY: MOBILE DEPLOYMENT

- **Objective 6: Assess capabilities to send MT63-2000L bulletins over VHF frequencies, and to receive and store them.**

  CAPABILITY: MT63 SKILLS

- **Objective 7:** **Assess capabilities of individual volunteers to participate in PACKET CHAT**

  CAPABILITY: PACKET CHAT

- **Objective 8:** **Assess the capability of LINBPQ packet chat functions to serve as many as 6 simultaneous roundtable discussants trying to determine the best solution to a communications problem.**

  CAPABILITY: LINBPQ CHAT FUNCTIONS

- **Objective 9:** **Provide an opportunity for participants to utilize ICS Forms 211 (Incident Check In Form) and 214 (Activity Log), using handwriting, and inside WINLINK, Form 213 (General Message Form), as well as refer to Form 205 (Incident Communications Plan) to facilitate communications.**

  CAPABILITY: ICS FORMS

Scenario Summary

Saturday, May 6 2017

Our scenario included 4 different time periods representing 3 days' scenario of a devastating hurricane that temporarily destroyed some or all of local conventional communications systems, including telephone and Internet.

Preparation	Initial Response	Continued Recovery	Continued Recovery	Cleanup
0830 Actual Time	0900 Actual Time	1000 Actual Time	1100 Actual Time	1200 Actual Time
Day 1, Noon, storm approaching	Day 1 2359, storm has passed	Day 2 Noon recovery	Day 2 2359 recovery	Day 3 Noon wrap-up

Participants in the Exercise were unaware of what communications difficulties would be simulated in each segment of the Exercise. At each time slot, they opened sealed envelopes and were given (in both hardcopy and flash drive format) the current situation and their goals for the segment.

SCHEDULE OF HANDICAPS:

	0830 Day 1 Noon	0900 Day 1 MN	1000 Day 2 Noon	1100 Day 2 MN	1200 Day 3 Noon
Universal Handicaps	No handicaps	Telephone / Internet Down Ham radio voice repeaters down.	Telephone / Internet Down Ham radio voice repeaters down.	Telephone / Internet Down. 2 Ham radio voice repeaters up: SARNET & 146.91 work	Everything works
EOC			Fixed antennas fail		
RED CROSS		Utility power out.	Utility power out, antennas down.	Utility power out.	
NEWBERRY		Fixed antennas fail.	Utility power out.	Utility power out.	
SR CENTER			Utility power out.		

Note: facilities which had to put up their own antennas from the very beginning did not have to take them down during "antenna out" periods; they were considered to have already passed that test.

At each time period, each location was assigned one or more messages to communicate. These were of several possible types, including short "Tactical" messages (a line or two of instructions that could be easily delivered by voice), ICS-213 "record" traffic (which needs to be communicated with word-perfect fidelity); email attachments such as a map or spreadsheet, or a discussion to be held over radio.

In addition, the EOC was suddenly tasked with a requirement to dispatch a mobile communications team to a nearby hospital to carry out a simulated emergency communications task related to an extreme emergency at the hospital (need to evacuate an entire building due to damage sustained).

At all times after the initial 30 minutes, teams at all four locations had no means to communicate with, or assess the situation of, any other location, other than by amateur radio communications – simulating what would happen if an actual telecommunications emergency were to occur.

SECTION 3: ANALYSIS OF CAPABILITIES

This section of the report reviews the performance of the exercised capabilities, activities, and tasks. In this section, observations are organized by capability and associated activities. The capabilities linked to the exercise objectives of Operation Hurricane Test are listed below, followed by corresponding activities. Each activity is followed by related observations, which include references, analysis, and recommendations.

CAPABILITY 1: ANTENNA PLACEMENT

Capability Summary: Fixed, pre-existing antennas should obviously perform for necessary communications. However, these may be damaged by high winds, and volunteers need to have the skills to efficiently replace them with ad-hoc created or installed antennas.

Activity 1.1: At some point, each location had to install new antennas.

Observation 1.1.1: Mixed. Our previous documentation has shown very significant deficiencies in fixed, pre-existing antennas at all locations. There are no fixed antennas at the Newberry Hurricane Shelter or the Senior Center. The antennas at the EOC have been shown inadequate for HF communications of all types, and inadequate for VHF communications should repeaters fail. ***During this exercise volunteers placed antennas at all sites, with general but not universal success:***

- EOC, Division of Forestry placed a mobile 50 foot tower, with VHF antennas at the top (success); HF antenna affixed to the tower however did not achieve successful communications (possibly due to operator inexperience with equipment).

- Senior Center: volunteers placed VHF and HF antennas with skillful use of lines placed over nearby pine tree limbs using a "Chuck-It" dog toy.

- Newberry: Volunteers placed VHF and HF antennas, using a rope to the foul-ball pole. They had difficulties getting their HF antenna to work, but generally succeeded with their VHF antenna.

- Red Cross: Utilized good fixed antenna and added ad-hoc vhf antennas as needed.   Did not place an HF antenna.

Analysis: Well-known limitations of the EOC and Newberry hurricane shelter antenna situations still remain to be addressed.   Volunteer skill at achieving height seems to have become significant, while there are still some technical gaps in the ability to couple power effectively to some antennas, as demonstrated by difficulties at the Newberry center.

REFERENCES:

Previous documentation of EOC antenna issues: http://qsl.net/kx4z/EOCTotalAntennaProposal.pdf

http://www.qsl.net/kx4z/MyNewVHFandHFproposal2.pdf

EOC "how-to" document on HF systems:

http://qsl.net/kx4z/EOCHFWinlinkExpressPrimer.pdf

Recommendations:

- **Raise EOC VHF amateur radio antennas to at least the 50 foot level on their existing tower, to match the level used by the Division of Forestry mobile tower. (Appendix A, #6)**

- **Even with a 50 foot tall antenna, the relatively low-lying EOC location was unable to reach the SARNET all-Florida UHF repeater system. They will require voice relay to reach that system (Appendix A, #22)**

- **Establish an EOC horizontal multi-band horizontal dipole. Route cable as needed to reach these antennas, and provide access for ad-hoc replacement should the fixed antennas be damaged during hurricanes. (Appendix A: #11)**

- **EOC needed additional "ladder line" transmission line on the provided emergency random length dipole (Appendix A, #17)**

- **Provide more training opportunities for EOC HF capabilities (Appendix A, #23)**

- **Install VHF antennas at the Easton-Newberry Sports Complex. Provide a pull rope to allow an HF wire antenna to be raised if needed. (Appendix A: #12)**

- **Provide a ladder for Easton-Newberry team (Appendix A, #19)**

- **Provide more chances for HF antenna training for Newberry volunteer personnel (Appendix A #15, #30)**

- **Install VHF antennas at the Senior Center (planned, but not completed)**

- **Install HF antenna at the Red Cross (no plans at present time)**

CAPABILITY 2: EMERGENCY SIMPLEX REPEATER

Capability Summary: If existing duplex amateur voice repeaters are overwhelmed, or out of service, a portable simplex repeater (that acts like a digital voice recorder, and replays over the air, from a high location, messages received) can provide voice coverage to a devastated area.

Activity 6.1 Emplace a portable simplex voice repeater and utilize when advantageous.

Observation 6.1.1: Mixed. Team leaders wisely placed this asset prior to the beginning of the test, taking advantage of high antennas in the W4DFU Dental Tower amateur radio station, which is available to several of the volunteers, and very amenable to emergency usage. However, there was both a transmitter power limitation (caused by a poorly chosen power cabling) and a receiver sensitivity issue).

Analysis Transmitter power issue is both easily fixed, and did not hamper outbound voice communications to Newberry, the farthest location. However, the receiver sensitivity issue could be a low-performance transceiver or a high-noise environment on the top of the dental tower (where an APRS system is co-existent).

Recommendation:

- **Fix the transmitter power cabling issue.**

- **Test an improved receiver to determine what is the exact cause of the sensitivity problem. (Appendix A, #7)**

CAPABILITY 3: WINLINK COMMUNICATIONS

Capability Summary: WINLINK provides a world-wide, radio-based email capability that has been leveraged by mariners, emergency communications personnel, missionaries, and the Federal Government. Allowing both email and attachments, it can speed digital messages toward areas where the Internet is still working, and then forward them by the far-faster internet email facilities, or in a complete national disaster, can slowly move email to "Message Pickup Stations" by radio alone. It is the premier HF radio-based email system in the world today.

Activity 3.1 Generate, forward, and retrieve multiple emails and attachments via WINLINK, either using HF, or VHF capabilities.

Observation 3.1.1: Strength. Three out of the 4 locations were successful at utilizing the capabilities of WINLINK to get communications accomplished when the voice repeater were simulated "down".

Analysis Our local ARES group has developed a significant strength in this area, following the lead of the Marion County Hospital Emergency Communications group.

Observation 3.1.2: Weaknesses. There were problems with the automated HF-based relay station's software. The Red Cross location was unable to develop digital messaging operational status.

Analysis All WINLINK software (both client and server) is free, volunteer-developed software, and although the system is now 2 decades old, updates and corrections appear almost weekly. Some of the hiccups of the automated relay station (KX4Z) were corrected with a software update (Appendix A, #1, #2, #3, #8), but it would be wise for more local amateurs to lose their dependence on that automated station by developing their own client HF station capabilities, following the example of the Marion County Hospital Emergency Communications group. Additionally, simple software mis-configurations caused the failure at the Red Cross, and were exacerbated by interpersonnel difficulties that can be avoided. Additional training has already corrected the software mis-configuration. (Appendix A, #4)

Recommendations:

- Strongly encourage all ARES personnel to develop WINLINK capabilities, monitor the capabilities of volunteers, and take their proven skills into account in assignments.

- Encourage client hf WINLINK skills (Appendix A, #18).

- Train on peer-to-peer WINLINK (Appendix A #26).

- Develop EASYTERM skills as local alternative to WINLINK email. (Appendix A, #14, #25)

CAPABILITY 4: BACKUP POWER

Capability Summary:

Electrical Utility power loss is one of the most frequent occurrences in hurricanes, and is a major cause of loss of traditional communications. Amateur radio emergency volunteers need to have alternate power capabilities.

Activity 4.1 At some point during the exercise, every location except the EOC (which has strong backup power facilities) was simulated "power down" and had to work off alternate power.

Observation 4.1.1: Strength. Every one of our power-down facilities had battery, generator, vehicular or other backup power source.

Analysis Considerable effort into this strength has borne results. It was practiced at the table-top exercise conducted just a few weeks before the Full Scale Exercise.

Recommendation:

- **Continue to develop strengths.**

CAPABILITY 5: MOBILE DEPLOYMENT

Capability Summary: In a true communications emergency, it is likely that there will be additional locations that suddenly develop an emergency need for communications. Amateur radio volunteers should maintain the ability to service those needs through mobile vehicles, possibly including dis-mountable VHF and HF gear that can be set up quickly at a new fixed site, including antennas.

Activity 5.1 An unexpected request was delivered to the EOC location to provide mobile emergency communications to a simulated official of the Shands Hospital.

> **Observation 5.1.1: Weakness.**   Due to inadequate available personnel, and the general crush of task assignments and relative skills available at the EOC, the team was unable to send a mobile team to the requested site.
>
> > **Analysis** Simply not enough volunteers.   We had reasonable participation from the existing registered ARES volunteers in Alachua County (55%) but we will need to develop an even larger number of trained, exercise-participating volunteers.

Recommendation:

Find ways to develop new volunteers.   Additional "marketing" through CERT and other training, and encouragement of local GARS , GARC and other club members to participate might help. (Appendix A, #10, #28)

CAPABILITY 6: MT63 SKILLS

Capability Summary:

MT63 is a fast digital keyboard-based and potentially file-based mechanism to send accurate broadcast (1-to-many) information that can be very effective in sending bullets and broadcast messages.

Activity 6.1 Utilize digital means to send broadcast bulletins to mutliple centers.

> **Observation 6.1.1: Weakness.**   Our teams just didn't try MT63 during this exercise. Instead, they utilized WINLINK and/or EASYTERM as alternatives.
>
> > **Analysis** This isn't all bad, in that we have multiple overlapping skill sets to accomplish the desired communication and the teams were specifically instructed that any means was fair game to get messages across.   However, their recent familiarity with WINLINK outweighed their more distant expertise with FLDIGI software that includes MT63 protocol.

Recommendation:

- Continue strengths in WINLINK.
- Develop additional strengths in EASYTERM
- If time permits, retrain users on FLDIGI – as more become HF capable, they may develop this skill outside of our ARES training anyway.

CAPABILITY 7: PACKET CHAT

Capability Summary: Packet Chat skills were hoped to provide a way for participants to allow multi-party typed (digital) discussion similar to what can happen on a voice radio frequency. These skills were tested by a small number of participants in Thursday evening packet roundtables associated with other ARES training nets, and were easily acquired by participants. However, the function itself on the digital repeaters was found to be easily overloaded, so the utility of this skill without higher speed "mesh" communications networks is questionable.

Activity 7.1 Carry out a packet roundtable chat discussion of alternatives for simulated hospital evacuation.

Observation 7.1.1: Weakness. The packet chat capabilities of the existing infrastructure aren't sufficient to carry out this Activity for more than 2 or 3 participants.

Analysis This weakness was known before the Exercise was carried out, but not at the time of development of the Exercise. It was somewhat awkwardly handled with suggestions to carry out the roundtable via simple voice communications. Due to other difficulties in other tasks, this wasn't well communicated during the Exercise and a good roundtable discussion simply never occurred over any medium during the Exercise.

Recommendation: During the next similar simulation, better explain the nature of the roundtable and its need and function, and offer better options to carry it out.

CAPABILITY 8: LINBPQ CHAT FUNCTIONS

Capability Summary: LINBPQ, the software employed in much of the digital infrastructure created in the amateur community locally in the last year, allows for a "roundtable" chat discussion, forwarding each person's typed comments to the others involved. Unfortunately, the limitations of 1200 Baud Packet AX.25 are that this is unwieldy for more than about 3 active participants. Although in the planning stages it was hoped to be a useful function, by the time the Exercise had arrived, it was already known that the technology has significant limitations and its use was not as strongly advocated, with alternatives over voice suggested.

Activity 8.1 Roundtable discussion of simulated issues at Shands Hospital.

Observation 8.1.1: Weakness. Both due to ICS-213 communications weaknesses at the Red Cross and the weakness of this technology itself, this roundtable was not accomplished.

> **Analysis** Unless our group is able to bring about higher-speed technologies such as "mesh" tcp/ip high speed connections, voice communications will be a more effective method to carry out any required roundtable discussions.

Recommendation:

Further investigation into mesh technologies.

CAPABILITY 9: ICS FORMS

Capability Summary: ARES volunteers have been becoming more accustomed to standard FEMA/ICS forms through efforts of Jeff Capehart at previous simulation events.   It is desirable that they be familiar with personnel log in forms, and essential that they be familiar with communications logs and message formats, particularly ICS-213 ("general message").

Activity 9.1 Utilize ICS-205 frequency chart, ICS personnel log in forms, ICS communications logs, and transfer ICS-213 record traffic.

> **Observation 9.1.1:** Strength.   Most facilities utilized the personnel, communications logs, and ICS-213 forms well.
>
> > **Analysis** WINLINK makes sending ICS-213 particularly easy for those who are facile with this software.

> **Observation 9.1.2: Weakness**. Personnel at the Red Cross improperly transmitted ICS-213 "record" traffic as a one-line summary due to lack of skill at digital communications.
>
> > **Analysis** This was a particularly important finding, a very significant legal difficulty. Education has already taken place in a group setting to explain how crucial it is that "record" traffic be transmitted word-perfectly. Further training of the individuals involved needs to take place to afford them better communications skills.

> **Observation 9.1.3: Weakness**. Some facilities did not stick to expected and published callsigns in the ICS-205, making contact via WINLINK more difficult.
>
> > **Analysis** Unable to send WINLINK email properly if intended recipients are using an unexpected callsign/email.. .

Recommendation:

- Group education on the importance of word-perfect transmission of "record" traffic (Appendix A, #5, accomplished).

- Individual remedial training on digital communications skills (not yet accomplished).

- Group education on sticking to published callsigns/email addresses (already accomplished, Appendix A #24)

ADDITIONAL ISSUES/RECOMMENDATIONS

- EOC needed additional transceivers to remain "accessible" on more frequencies simultaneously. Addressed in Appendix A, #9, and possible additional transceivers are being considered.

- Team Dynamics might have benefited from having a group meeting at the kickoff of the exercise. Recommend having a "breakfast" before the next Full Scale Exercise (Appendix A, #13)

- Volunteers at the Senior Center had a problem with people just walking in to see what they were doing. Recommend signs or caution tape next time. (Appendix A, #16)

- The EOC has no means for amateur stations to connect to ad-hoc antennas outside; cables through penetrations were not available and the EOC team had to work in a covered portico outside the building. Recommend providing cables through penetrations to avoid personnel being exposed to elements/interference/harm from being outside.

- Newberry volunteer participants were also unable to be protected inside due to a very high cost of having the building made available for emergency exercises such as this. Recommend Newberry finding a way to make the building available for responsible individuals to use for emergency preparedness exercises such as this, once a year, for 5-6 hours. (The Senior Center found a way, without any charge.) (Appendix A, #20)

- Net Control duties were a significant consumer of volunteer time. When the EOC passed control unexpectedly to Newberry, antenna preparations at Newberry were delayed. Recommendation: train additional net control operators, and carefully consider the consequences, duties etc of emergent replacement personnel when considered. (Appendix A, #21)

SECTION 4: CONCLUSION

Operation ARES HURRICANE TEST was conducted on May 6, 2017 to test Alachua County ARES capabilities to provide backup emergency communications to a small sliver of Alachua County communications needs.

This was undoubtedly the largest, most ambitious Full Scale Exercise ever carried out by Alachua County ARES, at least within known history.   A very wide array of communications skills were put to the test, including simplex VHF voice, simplex VHF repeater, duplex VHF repeater, HF WINLINK, VHF WINLINK, and keyboard and file-transfer VHF packet communications.   Multiple voice and digital (AX.25) repeaters were utilized.   Alachua County is blessed with more than 5 duplex voice repeaters, and we now have more than 7 AX.25 digital repeaters, and are reconnected to the state-wide SEDAN packet digital network.

These digital skills (WINLINK, packet, HF) are quite new to our local volunteers, and this exercise was a huge stretch for their skills, assets, and strategies.   Furthermore, we strenuously tested their abilities to emplace emergency antennas and provide alternative power --- all things that would be important in a real hurricane / communications emergency.

As expected, there were successes and failure.   Amazingly, at least 53 messages were successfully transacted – including some to neighboring counties, likely a first for our group as well.

One of our volunteers put our next tasks very eloquently, when Larry Rovak said we needed to just keep *"growing our infrastructure."*

We intend to do just that.

APPENDIX A: ISSUES NOTED / IMPROVEMENT PLAN

(updates ongoing at: http://qsl.net/nf4rc)

Updated as of June 11 2017

No.	Issue	Suggestion	Actual Action Taken
1	RMS_RELAY 3.0.37.0 will not even start without internet access		**RESOLVED** Upgraded to RMS_RELAY 3.0.38.0 which DOES start without internet. Does not need authorization code. Now works perfectly: a) boots even if internet down b) automatically switches to radio mode if internet lost c) automatically goes back to internet if it becomes available. Good job, Winlink development team!
2	RMS_RELAY 3.0.36.0 ceased actually transferring email	Upgrade	**RESOLVED**

	after about 9 messages transferred		Upgraded to RMS_RELAY 3.0.38.0 which did transfer the messages.
3	RMS_RELAY 3.0.38.0 crashed in early evening	Possibly due to erroneous listing of AC4QS as MPS?	**RESOLVED** Crash posted to WINLINK program development google group for analysis/fix There were no actions taken there. NOTE: Has not recurred. Software has operated for > 7 days without problems. Considered resolved.
4	Art & Cindy unable to make soundcards connect for digital at Red Cross	Gordon to meet with them and try to figure out.	**RESOLVED** FIXED! There were several issues with soundmodem configuration: a) needed "single channel output" b) needed "1200 baud AX.25 VHF" c) needed center frequency set back to 1700 Hz (had gotten bumped) Made a "cheat sheet " envelopefor them...need to make more for other people...
4A	Make cheatsheet soundmodem setup helps for other people to avoid problems.		**RESOLVED** DONE, and posted to the web. http://www.qsl.net/nf4rc/UnderstandingAudioChannelConfiguration.pdf
5	ICS-213 messages not being transferred word-perfect	Further training on this important issue will be held.	**RESOLVED** Further training on this was held at the ARES meeting on Wednesday, May

			10th, and everyone present indicated they understood the importance of word perfect transmission of record traffic.
6	Even with 50 foot tower, EOC unable to reach Senior Center & Newberry Sports complex		Un-resolved. Jeff Bielling says they are at the mercy of the Sherriff 's department as to when ham radio antennas at the EOC will be raised to 50 feet. We don't have a firm date on when VHF antennas at the Newberry shelter will be installed. Best solution remains digital connections, and slingshots to emplace temporary antennas. EOC volutneers should be ready to place antennas in the trees south of the facility.
7	Newberry sports complex could HEAR the simp reptr at Dental Tower, but couldn't SEND to it	Replace the Baofeng transceiver with a transceiver with better receiver sensitivity	
8	After about 11AM, the packet gateways at KX4Z started to fail at receiving emails	Not sure what caused this, no one was home to observe. Figure out a way to investigate this, or make a way to have that part of the system reboot on command??	Possibly related to the RMS-RELAY that got replaced. Testing not yet accomplished. May require human oversight at KX4Z.
9	EOC needed more VHF receivers--unable to monitor all necessary frequencies		Proposal for EOC to purchase additional transceivers is being considered. Otherwise, volunteers should bring additional system.
10	EOC couldn't service all their requests -- unable to dispatch team to Shands	We needed more volunteers at the EOC.	**RESOLVED** Volunteers from Red Cross will be moved to the EOC to increase their number.
11	EOC HF station simply	1. need observable wattage	PROGRESS:

	couldn't be heard, while HF from elsewhere was easily getting into KX4Z gateway on 80 meters	output in line at EOC 2. EOC crew needs more experience at HF 3. EOC crew needs more antenna experience 4 EOC crew needs to be able to rapidly test WINLINK automated connections as a way to know their output success and propagation success	1. Don't yet have a wattmeter 2. 7 hours of training completed on 6/10/2017 with significant improvement in performance on EOC gear 3. Four different antennas temporarily raised on 6/10/2017 with significant improvement in volunteer understanding 4. EOC crew got much better at moving to new frequencies, learned how to use the pushbutton frequency entry.
12	HF difficulties at newberry	Difficulty getting antenna to match with tuning system.	Consider additional practice/training with random length tuning system. PARTIALLY RESOLVED: One volunteer gained significant additional experience with automated matching systems 6/10/2017. Consider obtaining auto-matching system. .
13.	Have Breakfast FIRST	Probably better to have group meet together before start, and delay the start to allow better setup time/completion	**RESOLVED** Future plans will include a get-together meeting at the start.
14	Frustrating not to get acknowlledgement of receipt of winlink digital messages	? Consider read receipts? ? Consider YAPP for local transfer?	Further training in EASYTERM is planned....
15	Newberry building casts a huge radio shadow -- needed	1. Expected VHF antenna on top of building will go a long	

	way to move antennas well away from building	way. 2. Possible way to hang HF antenna from top of building to stretch outward might help 3. Finding a way to connect to the light pole rung might allow for an antenna a long way from the building -- but need considerable feed line available	
16	Senior Center -- people wandered into the building!	Need signs or caution tape to explain and protect	
17	EOC needed longer ladder line to reach their antenna	Get more ladder line	**RESOLVED** Gordon ordered additional ladder line. Now we have to remember this or station some there.
18	Troupe found 40 meter digital spots full of QRM	(FCC allows only narrow range) -- ***consider using other winlink gateways out of the auto-slice***	**RESOLVED** **Use other gateways out of the auto slice.**
19.	Newberry needed a LADDER		**RESOLVED** **next time recommend ladders for people who will need to work on higher antennas.**
20	Newberry -- stuck outside -- needed more time to seet up survival "field day" paraphenalia -- couldn't read monitor screens in sunlight	Delay start of radio portion Find some way to get inside or find cover -- that shed?	**RESOLVED** **1. Provide more defined "set up time" after the breakfast next time.** **2. Make presentation to authorities in Newberry to gain increased assistance.**
21	when net control duties passed to Newberry, it hampered their setup	Shortage of trained helpers. Ask if someone can realistically	PARTIALY RESOLVED: Traffic handling on ARES net practice

		take on a task. (KX4Z might could have done it)	began on 6/8/2017.
22	Note: EOC was able to reach the SARNET		(No action needed but note successful SARNET connection)
23	Even with 50 foot tower, the best simplex from EOC to Newberry was S0 not workable	Even with proposed EOC VHF antenna height improvement, repeaters of some sort are still needed to reach Newberry. Expect same problem to High Springs - harden repeaters (NEWB and 146.91 worked very well) - develop HF skills/assets	Considerable HF training was provided at the General/Extra Class course, but other issues not yet resolved. Additional HF Training (7 hours) completed on 6/10/2017
24	EOC & Senior Center used varying call signs on digital	Better advertisement of expected call signs, and better sticking to announced call signs.	**RESOLVED** Was discussed at the May 10th ARES meeting so that people will stick to published call signs a bit better.
25	Unfamiliarity with EASYTERM		Group requested more training on Easyterm. Jeff Capehart is scheduling this.
26	No training so far on peer to peer WINLINK		Group requested training on WINLINK peer to peer at May 10th meeting Jeff Capehart is scheduling this.
27.	Didn't need so many USB drives -- possibly more copies of same thing?	Discuss best solution	**RESOLVED** Agreed that we'll just send ONE usb drive to each center in future tests.
28	Could have used more participation from home ARES members to help relay	? Practice transferring ICS213 and ARRL Radiograms on Thursday night net??	Jeff Capehart is working up plans to practice message passing.

	messages		Message passing training STARTED on ARES net on 6/8/2017.
29	Make Eval sheets more understandable		Devote some training time to this aspect with volunteer evaluators before next Full Scale Exercise.
30	Antenna analyzer would have helped the Newberry crew	As part of increased HF emphasis, work on procuring more antenna analyzers	Training for Gen/Extra was held that includes training on tuning antennas....

APPENDIX B: LESSONS LEARNED

While the After Action Report/Improvement Plan includes recommendations which support development of specific post-exercise corrective actions, exercises may also reveal lessons learned which can be shared with the broader homeland security audience. Federal Emergency Management Agency (FEMA) maintains the *Lessons Learned Information Sharing* (LLIS.gov) system as a means of sharing post-exercise lessons learned with the emergency response community. This appendix provides jurisdictions and organizations with an opportunity to nominate lessons learned from exercises for sharing on *LLIS.gov*.

For reference, the following are the categories and definitions used in LLIS.gov:

- **Lesson Learned:** Knowledge and experience, positive or negative, derived from actual incidents, such as the 9/11 attacks and Hurricane Katrina, as well as those derived from observations and historical study of operations, training, and exercises.

- **Best Practices:** Exemplary, peer-validated techniques, procedures, good ideas, or solutions that work and are solidly grounded in actual operations, training, and exercise experience.

- **Good Stories:** Exemplary, but non-peer-validated, initiatives (implemented by various jurisdictions) that have shown success in their specific environments and that may provide useful information to other communities and organizations.

- **Practice Note:** A brief description of innovative practices, procedures, methods, programs, or tactics that an organization uses to adapt to changing conditions or to overcome an obstacle or challenge.

Exercise Lessons Learned

The following subject headings are lessons derived from the Alachua County, Florida FSE on May 6, 2017 that are proposed for inclusion in the Department of Homeland Security's Lessons Learned/Best Practices web portal, LLIS.gov:

- The importance of effective Amateur Radio antennas at the EOC and expected hurricane shelter sites cannot be overemphasized, if effective backup communications are to be provided. It is crucial that this well-known weakness be corrected.

- THE ASSISTANCE OF STATE, LOCAL, AND PRIVATE ENTITIES CONTRIBUTED GREATLY TO THE LEARNING OPPORTUNITIES AFFORDED BY OUR EXERCISE.

APPENDIX C: PARTICIPANT FEEDBACK SUMMARY

PARTICIPANT FEEDBACK FORM

(SUGGESTED FOR USE IN SUBSEQUENT EXERCISES)

Exercise Name: _____   Exercise Date: _____

Participant Name: _____   Title:_____

Agency:_____

Role: __Player __Observer __Facilitator __Evaluator

PART I: RECOMMENDATIONS AND CORRECTIVE ACTIONS

1. Based on the exercise today and the tasks identified, list the top 3 strengths and/or areas that need improvement.

_____

_____

_____

2. Is there anything you saw in the exercise that the evaluator(s) might not have been able to experience, observe, and record?

_____

_____

_____

_____

3. Identify the corrective actions that should be taken to address the issues identified above. For each corrective action, indicate if it is a high, medium, or low priority.

_____

_____

_____

_____

4. Describe the corrective actions that relate to your area of responsibility. Who should be assigned responsibility for each corrective action?

_____

_____

_____

_____

5. List the applicable equipment, training, policies, plans, and procedures that should be reviewed, revised, or developed. Indicate the priority level for each.

_____

_____

PART II – EXERCISE DESIGN AND CONDUCT: ASSESSMENT

Please rate, on a scale of 1 to 5, your overall assessment of the exercise relative to the statements provided below, with **1** indicating **strong disagreement** with the statement and **5** indicating **strong agreement.**

Table C.1: *Participant Assessment*

Assessment Factor	Strongly Disagree				Strongly Agree
a. The exercise was well structured and organized.	1	2	3	4	5
b. The exercise scenario was plausible and realistic.	1	2	3	4	5
c. The facilitator/controller(s) was knowledgeable about the area of play and kept the exercise on target.	1	2	3	4	5
d. The exercise documentation provided to assist in preparing for and participating in the exercise was useful.	1	2	3	4	5
e. Participation in the exercise was appropriate for someone in my position.	1	2	3	4	5
f. The participants included the right people in terms of level and mix of disciplines.	1	2	3	4	5
g. This exercise allowed my agency/jurisdiction to practice and improve priority capabilities.	1	2	3	4	5
h. After this exercise, I believe my agency/jurisdiction is better prepared to deal successfully with the scenario that was exercised.	1	2	3	4	5

PART III – PARTICIPANT FEEDBACK

Please provide any recommendations on how this exercise or future exercises could be improved or enhanced.

_____

_____

_____

_____

_____

_____

APPENDIX D: EXERCISE EVENTS SUMMARY TABLE

Table D.1: *Exercise Events Summary*

Date	Time	Scenario Event, Simulated Player Inject, Player Action	Event/Action
05/03/17	830	All systems operational, participants establish connections	Roll call. Confirm understanding of methods by which each site can be reached, through multiple methods.
05/03/17	900	Hurricane damage: loss of all normal communications. Voice duplex repeater failure. Request for roundtable discussion. Request for mobile comms team to be send to Shands Hospital. Red Cross: No power. Newberry: Antenna fail.	Re-establish communications by alternate methods, roll call, dispatch comms team, begin transferring assigned messages.
05/03/17	1000	EOC fixed antenna failure Red Cross antenna failure, power out. Newberry power out. Red Cross power out.	Re-establish communications by alternate methods, roll call, begin transferring assigned messages.
05/03/17	1100	Red Cross power out.	Re-establish communications by alternate methods, roll call, begin transferring assigned messages.

		Newberry power out. 146.91 and SARNET repeaters reactivate.	
05/03/17	1200	All systems working again.	Complete passage of any remaining messages, close down operation.

APPENDIX E: ACRONYMS

Acronym	Meaning
AAR	After Action Report
ALS	Advanced life support
CDC	Centers for Disease Control and Prevention
DHS	Department of Homeland Security
EDS	Emergency Dispensing Site
EMA	Emergency Management Agency
EMS	Emergency Medical Services
FEMA	Federal Emergency Management Agency
FOUO	For Official Use Only
FPC	Final Planning Conference
HF	High Frequency (shortwave)
HSEEP	Homeland Security Exercise and Evaluation Program
IAP	Incident Action Plan
IC	Incident Commander
ICS	Incident Command System
IC/UC	Incident Command/Unified Command
IPC	Initial Planning Conference
LLIS	Lessons Learned Information Sharing
MDPH	Massachusetts Department of Public Health
MEMA	Massachusetts Emergency Management Agency
MPC	Midterm Planning Conference
MRC	Medical Reserve Corps
MSEL	Master Scenario Events List
NIMS	National Incident Management System
POC	Point of contact

Acronym	Meaning
RSS	Receipt, Stage and Storage facility
SARNET	Statewide Amateur Radio Networking (a connected series of amateur radio repeaters)
SNS	Strategic National Stockpile
TCL	Target Capabilities List
UC	Unified Command
VHF	Very High Frequency (30-300 MHz)
WINLINK	A radio email system, see www.winlink.org